Global Warming Made Simple

An Interesting Topic For Our Times

John Andreadakis

authorHOUSE®

AuthorHouse™
1663 Liberty Drive
Bloomington, IN 47403
www.authorhouse.com
Phone: 1-800-839-8640

First published by AuthorHouse   08/29/2011

ISBN: 978-1-4634-2744-3 (sc)
ISBN: 978-1-4634-2743-6 (hc)
ISBN: 978-1-4634-2742-9 (ebk)

Library of Congress Control Number: 2011910949

Printed in the United States of America

Table of Contents

Introduction

An Interesting Topic for Our Times

Many are the people today who are talking about this topic of "Global Warming". A **former vice president** of our country is talking about it and he also received a great gift from an institution which is involved with the betterment of our world.

Our **President** also is speaking about it, saying that we need to avoid using fossil fuels as their use greatly contributes to the phenomenon of global warming. He is telling us that the prices on energy consumption will necessarily have to "skyrocket," and that it is something we need to go through before it is too late. Not just our country raises the flag of warning but many other respected nations as well.

We see **video clips** of young polar bears swimming in the open seas because polar caps are melting. Other clips are showing the glaciers crumbling and falling down into the sea as a proof about the same problem.

Companies and whole industries advertize and are telling us that they are going "Green". Their products are environmentally safe and their prices can be higher since they protect the environment.

We begin to recognize that global warming is a serious problem. From my point of view not much proof has been presented as not many scientists are studying the problem. Still others say that the papers which are supporting the global warming projections may be flawed. Indications of a serious problem need to be studied seriously, not only by all scientists, but also by all people, even individuals who are not scientists. If the theory behind this problem is flawed, then it is criminal to have risen to such high levels in the governments of many countries.

Also, if you noticed, they are trying very hard to educate our young children on the matter and they are doing it very effectively.

I am of the school that believes the smartest people do not always become scientists. What I suggest with this is that all people should get involved with the study of such a serious problem and contribute to its solution. When people put their heads together, the results are greater than the smartest person among them.

Because this particular problem will affect us all we need to do the best we can to protect each other. Children are not the only ones who need to be educated; we need to be educated on this topic as well.

I am not scientist or a climatologist; I am not a politician. The issue of global warming has become a serious matter for all human beings and is a matter which should not be debated alone by politicians and those who have political interests. As of yet, we are not in any imminent danger. Still, we have in this very country many places where some global warming would be desirable, let alone countries like Canada and areas like Scandinavia and Siberia. And, yes, there are many places where the climate is very warm, but by the same token, they always have been warm.

What is the reason this debate exists today among politicians? Since the danger clearly does not seem to be imminent, it has to be characterized as a theory requiring a further study. And before any decisions are made and controls put in place, an impartial and vigorous examination must be performed by impartial scientists who have the qualifications to express their views and opinions on the matter, and the reasons and the information be made public without delay.

Former Vice President Al Gore passionately sounds the horn of warning that the ice caps are melting, and because of it, many lands will be submerged. He may be sincere in his fears, but is he earnest when he invests his money in a Chicago venture which will regulate energy consumption for everyone, especially the people in this country? His reasoning suggests the United States has been using energy resources for so long that it is only fair to leave some resources to be used by other countries which have not used their share.

If energy is controlled the way they recommend, what will it do to your food supply? And if you do not have food to eat, what does it matter what kind of short life you are going to have?

He did claim in the past to have invented the Internet. Or maybe his real intention was to control the Internet if he could, in hopes to profit from his invention. The issue, though, is much larger than Gore who continues to be up to his neck in venture capital interests.

I will not include politics in this topic, because politics is harder to fathom and measure, not knowing what is the interest of a politician. The politician will tell you that he does not need to be a scientist or an economist because he can hire the best to tell him what he needs. How different is it to say, "I will pay the best economist and the best scientists to say what I want to hear?"

In some respect what this politician is saying is true. You can hire the best scientists and the best economists to give you the data to match your theory. A theory is what most of the politicians are after, one that best satisfies their agenda and their real intentions and is divisive.

We know the reputation of politicians. And, yes, among them there are some good ones. Certainly, a politician should not be characterized as bad simply because we disagree with him. But others who seem good, after awhile are proven to be otherwise. Unfortunately, around them there is another class of people called lobbyists or others admirers or wannabe politicians. This is where things become murky because of the corruption of money, and we become divided.

A young man goes to a university to become a scientist. He learns the principles required to be a good scientist and learns to use tools not available to the average person, mainly computers and other instruments. But does the degree of that scientist make him a good scientist? It does contribute to the making of a good scientist, but is not necessarily the only factor. In fact, a person can become a scientist even without a degree. Circumstances, observations, conflicts and problems can make a person an above average scientist on a particular issue. On the other hand, money and other interests can make a well-qualified scientist a bad scientist, even a criminal scientist.

Quickly, I will state my position and my qualifications on the matter to which I will speak and I will challenge all good scientists to scrutinize my approach and the conclusions I will present in this book.

I was born during the difficult times of World War II on a small island in the Aegean Sea. My hometown and my high school bear the name of the renowned mathematician Pythagoras. At the time when I was young, people in my town were a seafaring people, yet my family was mainly involved with farming. That alone produced in me an internal conflict which produced the need for finding an exit. Here I was from a very young age going up and down the countryside with my donkey, especially on the hot summer days, while the other kids were splashing in the clear waters of the sea. Temptation often made me ignore the consequences and occasionally I did disappear from my duties.

Although I was bad at spelling in the Greek language, I was one of the best in math, geometry and trigonometry, as if I were a true descendant of the old Pythagoras. Based on my mathematical qualifications, an uncle of mine sent me several university math and physics books a year or two before I finished high school, books which I quickly studied in their entirety. After high school there were no means for me to pursue a higher education so I took the opportunity to follow a navigating officer's career with the Greek merchant ships, starting as an apprentice officer, a career which at the same time gave me the exit I was looking for.

My military service was in the Greek Air Force where I was exposed to electronics. There I was mainly with other young people who had taken classes in an electronics school already. It was in those classes where I did so well that they were asking me what school I had gone to. When I told them that I did not go to any school of higher education, they were accusing me of lying.

Later, to become a navigating officer, I took several courses related to weather, electronic navigating instruments, languages and other topics. With the ships, I learned English and Spanish, languages to which I had no exposure before.

As a navigating officer I witnessed weather conditions and recorded in logbooks the weather patterns we encountered en route. For the safety of the ship and ourselves, we had to study and forecast the weather ahead. I traveled all the oceans several times and very often in specific seas. I witnessed personally tides of as much as 50 feet high or known currents of as much as 8 knots, storms, hurricanes and monsoons, and naturally, I often asked the question as to why these things were happening in some areas and why not in others. Then just being on relatively older ships, we often had the need to be involved with electronic equipment problems such as with radars and gyrocompasses, especially when you needed them away from port.

Seven years on the ships was enough for me so I immigrated to the United States to study more electronics. After graduating from a technical school, I was offered a job in a major computer company as an electronics engineer.

For this company and for their equipment I worked for almost 30 years and in the end of my career with the company, I studied heating and air conditioning. Many principles in this industry are related to this topic referring to hydrocarbons, humidity, temperature, energy, evaporation, and other conditions.

Are these experiences enough to qualify me to make statements related to global warming? I am absolutely sure that I am qualified and I challenge all of you to test my observations and my conclusions. The principles I will describe are not only for scientists to understand, so do not be afraid to try to understand them and make your own conclusions as to who is right and who is wrong. Some people try to change your life and the life of our children. So please give it a try. To understand the details and the aspects of the phenomena does require scientific knowledge and equipment but the basic principles of the phenomena are simple for anyone to understand.

Gulf Stream

This is what Wikipedia says about it:

The **Gulf Stream***, together with its northern extension towards Europe, The North Atlantic Drift, is a powerful, warm, and swift Atlantic ocean current that originates at the tip of Florida,* (Well, not really, this is simply where the phenomenon becomes visible. It originates in a vast area in the tropics extending from Africa to the Caribbean Sea. Because the current in this area is so wide, it appears as an unrelated oceanic current and is given a different name. Eventually, this current enters in the Gulf of Mexico, and once in there, having no place to go west, it turns east between Cuba and Florida where it meets another stream pushing west from the northern side of Cuba. At this point the only escape is to head north by the east coast of Florida. This is the area where the stream not only becomes visible but it also is the strongest.) *and follows the eastern coastlines of the United States and Newfoundland before crossing the Atlantic Ocean. The process of* **western intensification** (No matter what the name is, it describes the phenomenon where two spread out and slow currents merge into a narrow area. The speed of the combined current intensifies) *causes the Gulf Stream to be a* **northward accelerating current** (accelerating for the same reason) *off the east coast of North America. At about 40N 30W, it splits in two, with the northern stream crossing to northern Europe and the southern stream recirculating off West Africa. The Gulf Stream influences the climate of the east coast of North America from Florida to Newfoundland* (not really), *and the west coast of Europe* (that it does). *Although there has been recent debate* (in my opinion, there is nothing to debate on this issue)*, there is consensus that the climate of Western Europe and Northern Europe is warmer than it would otherwise be due to the North Atlantic drift, one of the branches **from the tail*** (there is no tail) *of the Gulf Stream. It is part of the North Atlantic* **Gyre** (If you give it a Greekish name, does it make it more scientific?). *Its presence has led to the development of strong cyclones of all types, both within the atmosphere and within the ocean* (This is completely inaccurate. The weather patterns are the ones which influence the stream and also generate the stream). *The Gulf Stream is also a significant potential source of renewable power generation.*

Some estimate that the Gulf Stream alone packs 100 times the energy consumed in the whole world. And this is not the only strong current in the Earth's oceans. There is a strong stream in the southern Atlantic between South America and South Africa. Another strong stream exists in the Indian Ocean and similar strong streams exist in the Pacific.

Observations and endless studies have been performed on these oceanic currents but very few if any come close to present the actual cause of the phenomenon. There is also a formation of an endless list of words and terms to describe these observations.

You can spend a large portion of your life reading the studies of universities and of private companies on oceanography. When you try to study them, you find that all they try to do is to observe the phenomenon and the ways with which they did the observation. Invariably, depending on their understanding, they develop terminology, and in so doing, they appear occasionally to contradict, and, of course, what they say appears to be confusing or without any special purpose. What they all are missing is the simple fact which is the real cause of these currents.

Niagara Falls

How many of you have stood by the falls? Watching the falls has become a profitable industry. A very small portion of the energy is being used to generate electricity, yet it is not hard to realize that almost all that water uncontrollably keeps running down day and night all year long, year after year, wasting valuable energy. I imagine for the local area it is more profitable to exploit the spectacle rather than use the energy which is provided to them freely. Now, can you imagine how many falls exist on the earth, many of which are even larger than the Niagara Falls?

Standing by the falls, it is easy to visualize the energy hidden under the water which runs down to lower levels of the earth. All rivers flow from somewhere in the mountains and no matter how slow they proceed like the Mississippi or fast like Niagara Falls, the amount of energy released is the same.

In many instances, people in our day close the down-flow of a river to form a dam which is then used to transform the energy to electrical energy and have this produce a considerable amount of renewable energy. When I was very young I remember us taking our grain to be milled in a place like this outside of our town. If you live next to a river or you are using the river for your boat, you have never asked how this water was channeled into the river. The existence of rivers has marked their place in human history for thousands of years. Can you imagine how much energy you need to fill these rivers with their water? Let's imagine that there was no rain to fill them up and you had to pump that water with electricity. What amount of electricity would we need to use to accomplish this task? Some will say, "Well, there is no need to use any energy because there is no energy in having snow and rain in the mountains." Let me assure you that there is real energy which is consumed in having snow and rain carried to the mountains and, indeed, lots of it. In fact, there is exactly the same amount of energy for the rain to take the water there as if we had to pump that water there. We just do not consider it because we do not have to do it. It is done freely for us.

Water

Lake Effect

Public Domain image of Lake Effect view from satellite
This is a NASA satellite image of the Great Lakes.

Both examples, the Gulf Stream and Niagara Falls, which I have mentioned above, are related to water. These examples use a different aspect of the characteristics water exhibits, but both are equally important and related.

The information I will provide in this section in layman's terms is commonly known and used by both science and technology.

Lately, due to the introduction of Doppler radar and new equipment in meteorology, we hear more frequently the term "**lake effect**". Some people who are not very familiar with the term sometimes feel annoyed with it, saying, "All of sudden, we have a new reason for having snow." With a little more attention and with the aid of the radar images, you can see what takes place in the case of "lake effect". And with a photo such as the one above, you can immortalize a lake effect view that you cannot get standing next to the lake.

In pictures like this it is easy to see how much moisture evaporates even in cold weather, and immediately it becomes visible, thanks to the very low temperature. These pictures also show that the evaporation is continuous. Some associate the term "Lake Effect" even in the scientific community just with snow. They do this because snow has its inherent problems, is visible, and being visible leads you to discover there is no snow on the other side of the lake. The important aspects of the phenomenon are these: You have snow because there is continuous evaporation over a body of any water, not just in a lake, but also at sea. Sometimes during the winter in Maine by the coast they have the exact same phenomenon. They do not call it "lake effect" of course. They call it "Nor'easter". But make no mistake the conditions are identical. You have continuous evaporation over the surface of water, any water, lake, sea or wet clothes on a clothesline.

Evaporation of water happens all the time and is the strongest when it is invisible, especially when the temperature is the highest. Evaporated moisture in the air, depending on the air temperature, can be invisible, or can become visible like a cloud, fog, rain or even snow. This is all it is. The result may seem dramatically different yet the phenomenon is always the same. I should add here that evaporation only stops when all the water has been evaporated or when a layer of ice has been formed over it. Even in the case of lake effect or Nor'easter the important thing to understand is that evaporation is continuous and lasting over water. If it was not so, the wind would take the moisture away and you would see a cloud going away and not replenished. Today is January 23, 2011 and it is one of the coldest days we have experienced here in Maine in many years. The weather report showed a similar picture of white stripes of white clouds, not above a lake but above the sea extending out into the ocean. The stripes, as in the picture above, are the proof that evaporation is continuous and on the whole surface of the water.

There is always evaporation that takes place on the open surface of water. It is the same reason wet cloths dry out when hung on the clothesline. But in the tropical open seas this evaporation per square foot is hundreds of times larger and yet is invisible. We know about lake effect today and talk about it because we have the means to see it. But for the real and enormous evaporation which takes effect in the open seas, there is not even a special word to describe its astronomical magnitude.

Fog, Cloud and Dew Point

Fog and clouds are really one and the same thing. Simply a fog is a cloud on the ground. Normally the air we breathe can contain a certain amount of moisture. The amount of moisture air can contain varies in accordance to the temperature. The higher the temperature, the larger the amount of water it can contain. For example, the same amount of air at saturation point is not the same when the temperature is 30 degrees or when it is 60 degrees. Eventually, the moisture in the air can reach the saturation point beyond which no evaporation can take place. When the air becomes saturated at a particular temperature, this is called the **dew point**. So when the temperature on the ground becomes the same as the dew point, then the fog appears. This is why when the sun comes out in the morning it burns off the fog patches quickly. It happens because even a small rise in the temperature above the dew point makes the presence of moisture in the air disappear. As the warm air rises, eventually it reaches to a point where the temperature is equal to the dew point, and then the moisture becomes visible as a cloud. If the temperature drops farther, the excess of moisture condenses to water (rain), which in lower temperatures eventually becomes snow.

Water Idiosynyrasies

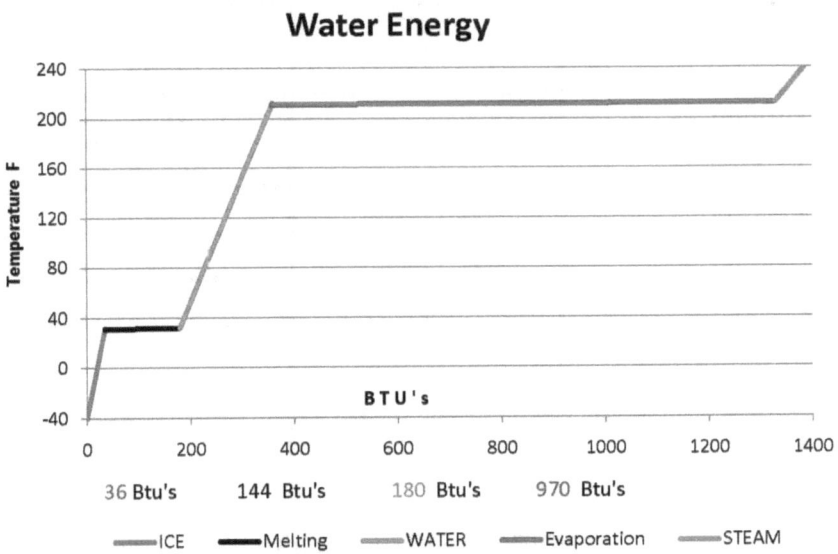

MATERIALS		BTU/lb/F
Wood		0.327
Water		1
Ice		0.504
Iron		0.129
Mercury		0.0333
Alcohol		0.615
Copper		0.095
Sulphur		0.177
Glass		0.187
Graphite		0.2
Brick		0.2
Glycerine		0.576
R-717	Refrigerant	1.1
R-744	Refrigerant	0.6
R-502	Refrigerant	0.256
Salt Brine	20%	0.85
R-12		0.213
R-22		0.26

Specific Heat Capacity

1. What I have described so far about water appears simple and normal. When we study water in a lab, it demonstrates some interesting peculiarities not readily observed when compared to other elements. These peculiarities are well-documented as is shown with the graph and the table above. The statements below serve to explain what the data means. The graph above demonstrates these peculiarities of water.

2. The energy needed to change the temperature of 1 lb of water 1 degree is called 1 BTU. To change the temperature of one lb of any matter one degree is specific and constant for that particular matter in a certain state. One lb of any item in the vast list of the different known elements only needs a portion of 1 BTU to change its temperature one degree. Very few elements need one BTU or slightly more than one BTU. A metallic element not only needs a small amount of energy to change its temperature, it also absorbs it easily. In contrast insulating materials resist absorbing any energy.

3. When water is ice it only needs **one-half BTU** to change the temperature of 1 lb of ice one degree. In other words, to change 1 lb of ice from -40^0 to ice +32^0, it needs **36 BTU.**

4. One lb of ice takes a larger space than one lb of water. This is what makes ice float, to the detriment of the Titanic. If ice was to sink in the oceans, then the conditions on our planet would have been completely different than what they are now.

5. At 32 degrees, ice melts to a liquid. The peculiarity here is that one lb of ice needs **144 BTU** to become liquid water at the same temperature of 32 degrees.

6. Changing liquid from 32 degrees up to the boiling point of 212 degrees needs one BTU per degree of temperature change for a total of **180 BTU.**

7. Finally here is the astonishing characteristic: Converting 1 lb of water at 212 degrees to 212 degrees of vapor needs **970 BTU.** Almost 1,000 BTU. This last peculiarity may seem untrue because you do not see the 212 degree temperature.

8. Evaporation, though, takes place at any temperature of liquid water. Boiling is the state where water or any other liquid is forced to evaporate meaning changing its state from a liquid state to a vapor state. The way we force liquids to evaporate (bring them to a boiling state) is performed in three different ways. One is to provide the energy to heat it; another is by providing the same amount of energy, reducing the pressure in its container as we do in refrigeration, and the third is relatively slow by maintaining the air above the water below saturation. The temperature at which a liquid can evaporate without providing heating depends on the pressure inside the container of the liquid.

You can easily demonstrate water boiling without providing any heat. If you have access to a vacuum pump, you fill a clear glass bottle halfway with water. Then you connect the pump to the bottle and you start pulling the air out. Within seconds you see the water boiling while the bottle becomes super cool to the touch. This is because the evaporation takes energy from the liquid and from the bottle. This is exactly what happens when a refrigerant is released inside the refrigerator or in the air conditioner.

These conclusions are scientific conclusions and are well-documented. Yet the effect of evaporation was well known without the science. I was a young lad on my island when I was spending my summer days working with my family in the fields. We were getting our water in a clay container. Then we were taking a wet towel and wrapping the container with the towel and placing it in a breezy shadowy place. You could easily tell the difference this produced to the water inside. You had nice cool water to drink.

Concerning water and its peculiarities, I need to clarify what science knows and what science does not know. Or what science is telling and what science is not telling us. With the help of a lab, science is attempting to measure and to prove a certain theory. With water they have studied thoroughly its characteristics or, in other words, irregularities. Raising the temperature of ice one degree is a consistent half BTU. And to go from -40 to +32 you need 36 BTU. To just change the state from 32 degree ice to 32 degree water you need 144 BTU and this is one irregularity. Going from 32 degrees water to 212 degrees water, the rate of change becomes 1 BTU per degree. The change in rate again is an irregularity but remains consistent in the whole water range. The biggest irregularity manifests itself in the evaporation. We need 360 BTU to go from -40 degree ice to 212 degree water, but just to change the state from water to vapor you need almost 1000 BTU, 970 to be exact. Scientists know very well these irregularities and have figured out how to use them. When I was young in my hometown in Greece they needed a lot of ice for their fishing boats, and just in our hometown, there were three ice-making factories. Were they scientists or inventors to have these factories? No, they just were able to find the knowhow from somewhere. Still there is a lot of information scientists do not know. They know that evaporation needs a lot of energy to take place. The clearest way to understand this is the following: If you want water to evaporate, you need somehow to provide the energy. When we boil water, you need to provide the electricity. But what happens with the lake? There is no free lunch. **The vapor steals the energy from the surface of water. In so doing, the water becomes cooler. It is simpler yet to say that the energy is hidden in the water vapor and that it defeats gravity.** When the vapor finds cooler air it gives the hidden energy to the cool air, becomes water and at the same time, loses its ability to defeat gravity. Scientists can see it happening and they know it, but they do not understand why it happens and this does not mean that I understand. It is the same in a lot of scientific work. If we knew how water defeats gravity with evaporation, we may be able to defeat gravity as well.

Glaciers

Glaciers are another related topic.

We spoke briefly about rain, snow and rivers.

Glaciers are formed when the accumulated snow exceeds the melted snow.

This is a true statement. And here is another true statement which is implied even if I did not point it out.

All snow comes from evaporation.

Yet where glaciers exist there is no water evaporation except for some very brief periods in the middle of the summer, and during that time, there is no snow as well. So where is the moisture which produces the snow coming from? We will talk about that.

Accumulated snow is compacted in layers and like iced rivers under the weight of the ice they move slowly to lower ground, forming a river of ice. Glaciers can be formed even in the tropics but only in high mountains. It takes many years for glaciers to travel their full length; sometimes it even takes hundreds of years. Ninety-nine percent of ice accumulated on the earth exists mainly at the poles of the earth. Winds and storms bring the moisture there, and because of the low temperatures, it becomes snow. Temperatures in the arctic areas are low because there is no direct sunlight. There are clear signs that glaciers existed in Europe and North America which have disappeared many centuries ago. We do not have a clear understanding of what caused these glaciers to disappear, nor do we definitely know what the form of the continents was at that time. One thing is certain, it was not because of our abuse of the environment as we are accused of doing nowadays. Also the slow flow of the existing glaciers implies that there was a continuous supply of snow over many years, even centuries.

When the glacier reaches the seawater it melts abruptly and produces the known spectacle of ice breaking and falling into the sea to form the icebergs. Ice melts at the sea because of the salty content of the sea. **Ice breaking from a glacier by the sea is not an indication of global warming; it is simply a natural phenomenon.** The reason glaciers retreat or expand in the vicinity of sea water is not an indication that global temperatures are rising today, but rather because snow accumulations have varied not today but rather hundreds or even thousands of years ago. The supply of snow was reduced not today but rather maybe many centuries ago. Therefore, the proper question to ask is not weather our climate is changing today, nor if the climate has changed 1000 years ago. **The proper question is how are snow and ice formed in the first place? Once you ask the right question, it does not take a genius to even answer the relative question of how does water fill the rivers and the waterfalls and the lakes up high away from sea level.**

The first thing in answering this important question is the realization that **clouds carry water** and snow to the higher elevations and the poles. "Clouds carry water"? Can you believe it? I made the same mistake. **Clouds do not carry water; clouds are water and in fact are only a small portion of the moisture in the air.** This is not the full answer, but is the correct and simple answer everyone can understand, and this leads you to the next correct question. **How are the clouds formed?** And another good question to ask is: **What brings the clouds to the mountains and to the poles?** For the casual observer, clouds appear to travel in different directions and at different heights and they have different shapes, even colors and scientific names. We have seen clouds from above and they are always cotton white. But from the ground they are only sometimes white. Most of the time clouds are dark gray. When they are dark sometimes they drop **rain**, or **snow** or even **hail**. Oh, there is something else that clouds do. They give the ominous display of power and energy called **lightning**. In Greece and in Maine I have seen trees blown to smithereens in a fraction of a second like they have been exploded from within. And in Colorado or Utah you may see fires started from lightning. But can you imagine the benefit we can have if someday we can harness the power clouds can give? Electricity and electrical science tells us that lightning is a wasted energy of tremendous proportions which is wasted much the same as the unused energy from a waterfall. This is basically all that a casual observer can deduct from observing clouds.

Doppler radar and lake effect gives us the next level of visual observation that clouds do form above lakes and that the **wind** does carry them to the land and that the clouds then can produce rain and snow which falls on the ground. But without the aid of the radar or science you cannot see how the cloud is formed above the lake. The first simple understanding science is giving us is that the moisture above the open surface of the lake is carried up and away and the lower temperatures at higher altitudes eventually make this moisture in the air visible in the form of a cloud. In this step, science tells us that **water evaporation** and the temperature of the wind is what generates and moves the **clouds** which produce **precipitation**. To make things clear, I repeat, clouds do not bring water but they <u>are</u> water and they represent only a small visible portion of moisture in the air. Moisture in the air cannot become rain or snow unless it first goes through that critical point (the dew point) which is the temperature that makes moisture visible.

Although glaciers are spectacular, in a similar fashion as the waterfalls, their origin is not any different from the origin of a river. The truth is that moisture has traveled this far and this high for them to exist. Their existence is a testament that their supply has not been interrupted through the whole span of their existence. As for the moisture the sun provided all the energy to cause the evaporation and the sun was also responsible for the wind which carried the moisture this far and this high. The statement I want to make here again in case you missed it earlier is this:

All rain and snow are the result of evaporation of water and practically the bulk of that takes place in the tropics.

An Important Calculation

There are statistics which indicate the average rainfall on every place of the earth daily. Few isolated areas, away from the sea (deserts), receive hardly any moisture at all, but, all the areas around the tropics up to 30 degrees latitude receive about 10 mm of rain daily. If we were to calculate how much water in pounds (lbs) falls on the earth, then we can calculate how much energy is needed to evaporate all this water. We do not need to be very accurate to figure this out. We only are interested in getting an idea of the magnitude. Very conservatively, we can assume that the average rainfall is not 10mm but only 2mm. We have the size of the surface of the earth in square kilometers (km).

Our first step will be, therefore, to calculate how much water each square meter receives with a average daily rainfall of 2mm. Then we will find the amount for a square kilometer, then for the whole earth.

The values we start with are the following:

Earth's surface: 500 Million Square Kilometers = 5×10^8 sq km
Daily average rainfall: 2 mm
1 lb of water is 0.5 liters   and 2 lbs is one liter.
One liter is 1,000,000 cubic mm or 2 lbs of water.
One square meter is 1,000,000 square mm.
One square meter 1mm high is 1,000,000 cubic mm and is still 2 lbs of water.
One square meter 2mm high is 2,000,000 cubic mm and it is 4 lbs of water.
One square meter receives 4 lbs of rain daily.
One square kilometer is 1,000,000 square meters so it receives 4,000,000 lbs of water daily.

Or every square kilometer on the average receives 4,000,000 lbs of water daily from rainfall.

Total daily rainfall on the earth is estimated to be:

500,000,000 x 4,000,000 = 2,000,000,000,000,000 = **2 x 10^{15} lbs of water**

Since all the rain and snow comes from evaporation and each lb of water needs approximately 1,000 BTU for evaporation, the total energy needed to evaporate the daily rainfall is: $2 \times 10^{15} \times 1,000 =$ **2 x 10^{18} BTU daily.**

1 Exajoule = 10^{15} BTU

Evaporation energy is 2 x 10^{18} BTU or approximately 2,000 Exajoules per day.

2,000 Exajoules/day is equivalent to 4.3 times the world's energy consumption in the entire year of 2008 (474 Exajoules for the entire year), or 2,000 x 365 days in a year makes 730,000 Exajoules of evaporation energy which is 1,540 times the world's annual energy consumption. Even this amount of energy is staggering compared with the amount of energy we consume in the whole earth and it shows that the sun is in charge, not us.

Do you grasp what is said here? By means of evaporation, water absorbs a portion of the energy coming to us from the sun. And this amount of energy equals 1,540 times the total amount of energy we consume in the whole earth. Please pay attention because I know that you still did not get it.

What I mean is this: **From all the energy we receive from the sun, a portion of it which equals 1,500 times the world's energy consumption is absorbed from the evaporation of water to moderate the heat on the earth's tropical waters.**

But rainfall in actuality is not all the energy involved to evaporate this amount of water which eventually will become rainfall. More energy from the sun is used to further warm up the oceans, especially in the tropics, despite the cooling effect evaporation has on the ocean water. Another very large portion of sun's energy is involved to further heat up the atmospheric air, especially in the tropics which generates the winds and storms that carry this moisture far and high. A lot of this warmer and moist air from the tropics finds its way to the mountains and to higher latitudes, even to the poles. Condensed evaporated water releases all the heat energy it absorbed from the ocean. A portion of this energy is radiated to open space and the rest becomes mechanical energy following the rainwater on its way down through the rivers back to the sea.

Cooling and Evaporation

Here I will describe a large air-conditioning unit and will compare its parts to our environment.

A large vacuum pump is using electrical energy to pull the refrigerant (freon gas) from the chamber of the machine which produces the cooling. When cooling is needed, freon which has been compressed by the same pump is released in the evacuated chamber of the unit. Because of the low pressure, the freon evaporates into gas and that absorbs heat from the walls of the evaporation chamber in the unit and drops the temperature of the walls of the chamber. Outside the walls of the chamber, a fan circulates the room air through it to cool the air. The evaporated gas absorbs heat from the air-conditioning unit and brings it out to the cooling tower. Large buildings make use of cooling towers, not to cool the interior of the building, but to strip the heat from the condenser, and at the same time to prepare the refrigerant to be reused for cooling. The big compressor pump pulls the gas from the evaporation chamber and compresses it in the condensing area in the cooling tower. The high pressure inside the condenser makes the gas become a liquid again and that condensing action releases the heat which the evaporation absorbed from the actual interior unit. The condenser then becomes hot because condensing of the gas has the reverse effect of what evaporation is doing. Another common name of an air conditioner is "heat pump" because this is what it really does. It pumps the heating energy from a closed area and releases it outside. Some units in actuality reverse their operation between condenser and evaporator on cold days and the unit still is able to squeeze heat out from the cold air outside and bring warmth inside the house or a room. In a large unit the condenser is located in a tower where water is used to cool it down. The condenser there is so hot the water which runs over the condenser boils to a visible white cloud coming out in the air from the cooling tower. Water evaporation in the cooling tower is the means of absorbing the heat from the condenser.

In the earth's environment the evaporation area is the surface of the ocean or the lake. Water evaporation there absorbs a lot of heat from the ocean or the lake. When moisture in the air condenses in high altitude it releases that energy which is radiated out into the open space as energy going away from the earth. Water is the refrigerant in this case. The pump is the heating energy from the sun.

What is more striking is the fact that evaporation absorbs energy from water and in so doing it prevents lake water or sea water from overheating. This is why I used the example of an air-conditioning unit here instead of a furnace. Let me recap, therefore, what is taking place:

The energy from the sun in the form of heating energy strikes the surface of the ocean. A portion of the energy received from the sun causes the evaporation which has the astronomical value calculated above. Another even larger portion of the sun's energy still continues to increase the temperature of the sea water further and also is used to heat and to lift the moist air mostly from around the equator. This lifting of the moist warm air around the equator is the engine which produces the **winds** and which is completely independent from the energy consumed for the evaporation. Later in this book we will discuss wind formation and the effects winds have on the environment.

Global Warming

The moral of the situation as it concerns global warming is this: We are concerned that our way of life may be contributing to global warming by one or two degrees. But where are we doing this? Obviously, if we do this, we do it in the cities which really are isolated areas on the land where the factories exist and where the bulk of our energy consumption is concentrated.

But even in these areas, when did you pass next to a building or a factory and you felt heat coming at you from it?

Once I worked on a summer job in a cement factory in Buffington, Indiana. Often when I needed to find relief from the sun, I would go under the big rotating kiln through which literally fire was going. In contrast where did you stand in an open place where you were not able to feel the warm rays from the sun? You feel it in Miami where, if you are walking down a sunny street, you tend to hesitate in the shade of a tree and take a deep breath before you walk out again into the open sun.

You feel it in the cool mountains of Colorado. There are many of you who went skiing on the freezing slopes of Colorado and you returned home with sunburn. I have been up on the top of Pikes Peak in Colorado often. Up there the temperature is very low even in the middle of the summer. If you stay motionless for a short time, you feel the side of your body in the direction of the sun getting hot and the other side you feel it freezing.

How can you compare the puny heat emission even under the isolated cement kiln with the rays of the sun which are felt in every inch of the whole earth? And there is even more unappreciated relief from the sun's rays when the moisture in the air up high in the sky becomes rain or snow.

Some will say the problem really is because we have too many cars. But even in this case, when did you stand next to your car and feel the heat coming from the engine of the car? You can feel it only when you touch the hood. We all have been in a car traveling and been sweating bullets sometimes, and we are lucky if we have air conditioning. But where does the heat in the car come from? It comes from the roof or from the window. It comes from the sun! Even in the middle of the winter you feel the warmth of the sun inside your car.

Here I am right now in Maine during the winter and every day I hope that the sun will be out again. Even with a clear sky we feel lucky when the temperatures rise above freezing into the thirties. Who in Scandinavia and in Siberia and in Alaska does not wish the winters to become shorter? The Scandinavians travel every summer to my island in Greece where I grew up and they even throw off their swimming clothes to soak up as much sun as they can. However, they also go under the umbrellas to avoid getting a painful sunburn.

When you see a cloud, please remember what these clouds mean for us:

First, when the moisture was released from water an enormous amount of heating energy was taken from that water and inappreciably we do not even know it.

Second, clouds are nothing more than a proof of moisture's existence in the air, and moisture with the help of the sun again produces the winds which carry moisture to our gardens and to our livestock to give us food. How could there be life without food? This is somehow appreciated by a lot of people.

Third, when moisture in the air condenses around the cloud, this is the point where the heat energy absorbed from the water down on earth is being released out away from the earth. This is also an unknown fact, and, as such, is unappreciated. Our planet is the perfect cooling tower.

A cooling tower on the Empire State Building provides cooling inside the building but releases a lot of heat to the outside air just when the utility company complains the most for the energy consumption spikes. And yet, even in the cooling tower, there is the benefit of water because when it evaporates in the tower, it takes all of the objectionable heating energy up to the sky by means of evaporation.

Let me ask you at this point - Am I convincing you with what I say here? If I am not yet, please continue because the following is a convincing proof of what role water and moisture play in our lives.

In my thirties and early forties I lived mostly in Indiana and Illinois and Massachusetts where humidity is rather high. Occasionally, I would go outside to do a little work in the yard and within ten minutes, I would give up because I was melting in my sweat in the summer, and freezing painfully during the winter. Not knowing clearly the problem then, I was attributing it to getting older.

When I was forty-three I was transferred to Colorado Springs where I decided to do the landscaping at my new house on my own. I surprised myself when I realized that I was able to stay busy in the yard all day long on hot summer days.

If you are not familiar with Colorado, Colorado Springs is just east of Pikes Peak Mountain. As the weather systems move from west to east in that latitude when air passes from the top of that mountain, it is getting stripped from its moisture and as the clouds come down on the east side of Pikes Peak, they eventually become warmer in the lower altitude and disappear. We also often have the phenomenon where rain continues after crossing to the east side but before it reaches the ground, it evaporates. As a result, air in Colorado Springs is very dry.

Occasionally, we would go one hour north to Denver or one hour south to Pueblo. Especially on summer days, I could not wait to return home, feeling uncomfortable like I was not able to breathe. Humidity in both of these places, although not very high is definitely a lot higher than in Colorado Springs because they are located close to mountain passes which invariably bring down a lot more moisture.

It was at that time when I learned what was wrong with me.

The human body is a perfect machine more than we appreciate most of the time. Breathing and sweating is what regulates the temperature in our body. Remember the temperature inside our body is 98.6 degrees. Yet we feel comfortable in an environment less than 70 degrees. In a good factory you find a large tall chimney. As in the fireplace or your furnace you need the CO_1 and the smoke to go out of your house. Well, there are other things like CO_2 which are going out, but who cares? However, CO_1 is a poison. It will kill you if you keep it inside your house. If you noticed not many complain about CO_1 much. I am sure former Vice President Al Gore, John Kerry, and John McCain have plenty of fireplaces in their multiple mansions. But the harmless CO_2 has been declared as a pollutant. I am sorry, but never in my exposure to scientific things have I encounter anything related to CO_2 having any effect related to energy.

Well, back to our body as a perfect machine. It has a chimney and sewer to discharge unneeded and harmful materials and some chemicals which are needed to process our food so that we can absorb the good nutrients. Our chimney is our nose which like in a good furnace you have intake of air to give you the oxygen you need for the internal combustion in your body. From your chimney nose out comes CO_2 to prove that you are alive. Release of moisture from our lungs serves to cool down our internal combustion engine. This is not the only way our body releases moisture. The skin on our entire body has little openings from where moisture and other unneeded things come out. Also things for protection of the skin come out such as salt. But for the purpose of our discussion, moisture released inside our lungs and from our skin keeps our body at a comfortable temperature.

Our body is almost perfect. When our body is getting hot because we are actively doing something, we start to sweat. We feel it and we realize that we need to do something more. What we do is throw off our clothes which allows air to get to our skin and evaporate the moisture from the sweat, and most of the time we are doing this without thinking. Most of us say: "If it is hot, throw off your clothes." Where in actuality we throw off our clothes to let the air get to our skin for that extra moisture to evaporate and make us quickly comfortable.

We also can see it with the animals. An elephant takes water and sprays his body. How about your dog? He cannot throw off his hide. What he does is open his mouth, sticks his tongue out long and starts panting to cool his body within. Are these animals scientists? They know what is more comfortable and that is cooling from evaporated moisture.

If this has not convinced you yet, wait, I have something else.

There are many places around the world like in Colorado Springs where the humidity content of the air is very low. Colorado Springs is fine because there you get water from the mountains and sometimes moisture from the nearby mountains in the form of rain and snow, even hail on summer afternoons.

There are other places though where there are not any sources of moisture nearby. Such areas usually are away from the sea or other bodies of water. Yes, the areas I am talking about are the deserts. In areas where the air has some moisture, the average temperature of the air rises about ten degrees during the day and releases about the same amount of heat during the night. In the desert where there is no humidity, temperatures rise quickly above a hundred and at night they drop below freezing. What does this mean?

It means this: All places receive the same amount of energy from the sun for the same latitude. Water and humidity in the air have this immeasurable effect on our environment. There is much more water in earth below sea level than there is dry land above sea level. This is enormous, and indeed, has an almost unlimited influence on our environment. When we are talking about water or humidity or moisture or clouds or snow and ice, all these are one and the same thing. It is water, H_2O. Dry land is less than one-third of the earth's surface, the rest being water. Let us suppose now that water is one-third of the earth's surface. This would cause most of the land to be like the Sahara Desert because there would not be enough moisture to moderate the climate.

I do not know everything, I admit. We know what CO_1 is and we know how to protect ourselves from it. We do it when we are not running the car engine in a closed garage and when we properly vent our furnaces and our stoves and our fireplaces. But I admit I do not know much about CO_2. Usually when scientists discover things, eventually they make these things known. What is the amount of CO_2 on the earth and in the air? What is the influence it has on the environment? And if it does, how does it do it? Why is there so much warning about global warming in the media and in our kids' schools, but why is there silence about the facts concerning CO_2?

Wind and Oceanic Currents

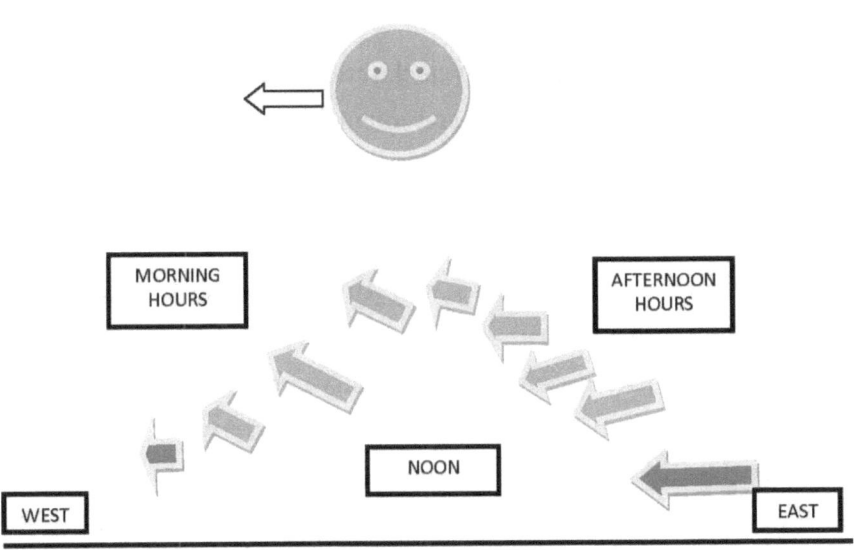

Winds are thoroughly studied and measured. In this extended study there is a detailed description of winds and good conclusions have been drawn as to how we can forecast them based on barometric pressures and prevailing temperatures around a certain location. This is the work meteorologists do to help us know how to get dressed, to travel and do a variety of outdoor activities. I suspect because of all the work performed and all the electronic equipment available in the industry, it develops the impression for everyone that there is no need for further study or conclusions.

The statement I will make here is not well known throughout the whole industry or known at all. But let me make the statement:

The prevailing wind direction under the sun is from east to west. Under the sun refers not to the whole earth but rather where the sun at noon is directly over our head. These are the areas between the solstices, or in other words, the tropics.

How can anyone make this declaration unless he personally has observed it? Well, I make this statement because I have observed it and I have contemplated what the ramifications of the statement are.

The above sketch explains why this is the case. In the early morning hours there is still a residual wind from east to west. As the sun rises the air increases in both humidity content and temperature, and begins to expand while it maintains its westward progression. By noon the air below the sun reaches its highest expansion upward, its highest temperature and it has absorbed a lot more moisture. In the area between 10 degrees north to 10 degrees south from the sun's track, the ascension of the air is the highest, yet is completely invisible because the air never reaches a dew point to become visible as a cloud. As the sun departs west, the air at the highest point begins to get cooler, increases in weight and begins to descend. The upward expanding lighter air develops a small vacuum below, so the descending cooler air increases the forward speed to fill that vacuum. Because the westward progress of the sun is at 900 nautical miles per hour, the air behind the sun attempts to fill the vacuum under the sun by increasing its speed but never reaches it. Nonetheless, this is repeated every day and in this way the westward movement of the air (wind) is maintained.

These persistent winds from east to west in the Atlantic tropics are responsible for the development of the surface currents in the same area and having the same direction as the wind. These are the Equatorial Currents. These same kinds of winds are responsible for the formation of all the Equatorial Currents in all the oceanic tropics. When these currents reach Central America, they have no other place to go and turn north and make the **Gulf Stream** visible as these surface currents combine in a very small area. It is simple. And I would not know that winds in the Caribbean Sea have an east to west direction if I had not traveled that area extensively. The truth is that it has been the perfect secret.

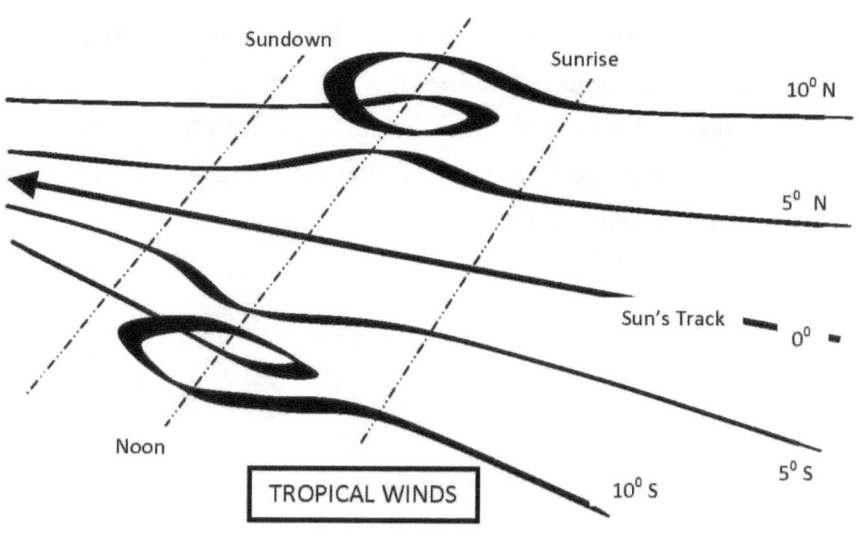

TROPICAL WINDS

Line "0⁰" is what we described in the previous diagram and it is directly under the sun. Because the view is directly from above we do not see the upward expansion of the air. In the areas 5 degrees north and south from the sun's track the air expands likewise as in area "0⁰" but it also expands away from the sun sideways and this is shown with the lines "5⁰". This diagram more clearly shows what happens in the areas 10 degrees north and south from the sun's track. In these areas the air represented with the lines "10⁰" still expands high much the same as it does in the air represented by the lines "0⁰" and "5⁰", but it also expands away from the sun's track. Here though, the hot humid air reaches the dew point and not only becomes visible forming clouds but also generates precipitation which enhances the downward drafts. These downward drafts still seek the lower barometric pressure developed under the sun. This cold air saturated with moisture is much heavier and under the influence of the falling rain sneaks under the warm expanding air marked with the lines "5⁰" turning in an eastward direction. Once this happens the tropical storm is formed and is invigorated as the sun passes every day over it. From then on these storms disintegrate only when they reach a land mass where they lose the supply of moisture they are getting from the warm surface water of the tropical seas.

The barometric pressure in the center of the tropical storm is extremely low, but, let me make it clear, this is not what is causing and sustaining the storm. It is actually the result of the tropical storm mechanics. Warm air with humidity in it is very light, and this is why hot air balloons do not fly well in moist warm air. This is why Colorado Springs and, in general, all the east side of the Rockies is the ideal place for them to fly. A collision of warm moist air with cold air is always violent because the condensation of the moisture way up in the sky instantly regains its gravity in the form of falling rain. Rain then forms the downdrafts and the downdrafts pull more cold air to mix with the humid air resulting in even more precipitation and strong winds. The violence then is even stronger when hail is introduced from the presence of even colder air. It is the same scenario which produces the strong tornados in the vicinity of a cold weather front.

Tropical storms are more pronounced, in frequency and in strength, right after the sun has reached the highest latitude and begins to return toward the equator. This is the time when the surface water in the tropics is the warmest.

This diagram also indicates why the wind rotation around a tropical storm is counterclockwise in the northern hemisphere and clockwise in the southern hemisphere. Tropical storms are formed and sustained for many days because there are two well-defined areas, one with warm humid air closer to the sun and the other away from the equator consisting of colder dry air. This scenario is the same as a stationary cold front located some 10 degrees away from the sun's track and some 30 degrees away from the equator. The initial formation of a downdraft has a westerly direction because the sun is pushing it this way, but then it turns toward the equator, because this is where a vacuum is formed when the moist air heated by the sun is moving upward in the sky. From then on all storms are the same. The winds in the sun's direction become warmer and rise while on the opposite side of the storm the winds become colder and invigorated as they plunge closer to the ground. In general the whole air mass in the tropics is moving from east to west because the sun is pushing it in that direction. One way or another this air mass from the tropics spills out in the direction of the poles, and in higher latitudes the air mass goes in the opposite direction from west to east and eventually circulates down to rejoin the tropical air mass. This is why tropical storms travel westward and high latitude storms travel the other way from west to east.

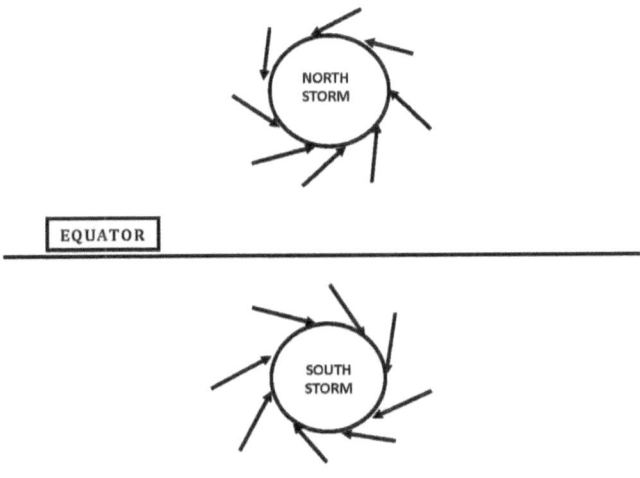

TROPICAL STORMS

All storms in the northern hemisphere differ from all the storms in the southern hemisphere in the direction their winds rotate around the barometric depression. Winds in the northern hemisphere rotate counter clockwise while in the southern hemisphere rotate clockwise.

At this point we will turn our attention more towards the influence the winds have over the surface currents of the oceans. Because the currents indicated on this map are accurate I wanted to show how the winds are related to the formation of these currents and clarify better the designations in this map about warm and cold currents.

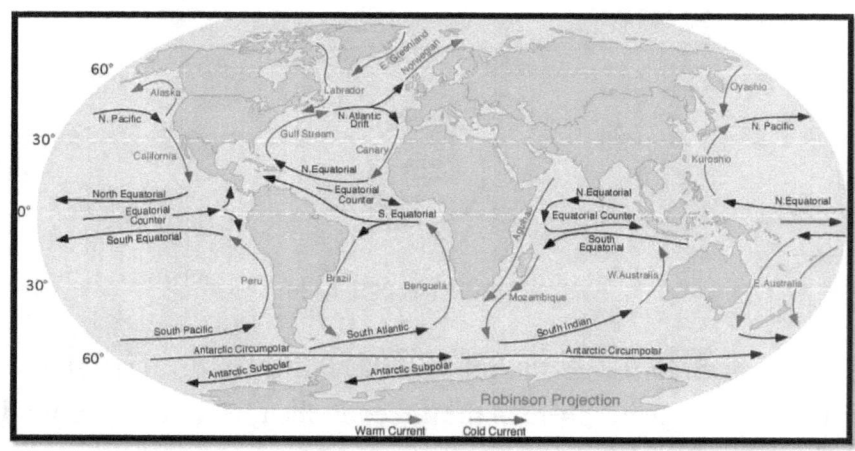

OCEANIC CURRENTS

In studying the oceanographic currents you find out that they are also well-documented because their documentation becomes very useful in sea navigation. Often I have been surprised at not finding anywhere an assertive explanation on what causes these currents. And whoever attempts to give an explanation is usually wrong as I stated at the beginning of the book. The closest I have seen in providing a correct explanation is that winds do cause shifting in these currents, but they fail to say that winds are causing the currents. This map above is copyrighted by the United States Government and it is very accurate in showing the oceanic currents and their names. For example in the Atlantic Ocean the North Equatorial Current appears to be extended to form the Gulf Stream then into the North Atlantic Drift. The North Atlantic Drift has to split in two because sea currents cannot stop and disappear abruptly. A portion of the current goes north to form the Norwegian Current but the biggest portion turns south to form the Canary Current. Then the Canary Current turns into the North Equatorial Current. Because these currents change names does not mean that the reason of their existence has changed. My contention at this point is that despite the different names these currents which circulate around the North Atlantic are one and the same current. The reason for their existence is only one as you will see. Now we will turn our attention to the South Equatorial Current as it comes from West Africa. Close to Brazil it has to split again in two. The south branch forms the Brazil Current and other follows the northern Brazilian coast. **Here in these tropical waters north of Brazil and in the Caribbean Sea is where you will find the engine which produces the currents in the North Atlantic Ocean.** This engine is none else but the perpetual winds blowing all year long from an east to west direction. One more detail that I want to add in this area of the map is this: The northern portion of the

South Equatorial Current does not just disappear. By the time both North and South Equatorial Currents reach the coast of Florida, they combine their forces to form the well-intensified current we call the Gulf Stream. From that point on there is nothing else which would intensify this current.

Looking on the global map above, there is somehow misleading information indicated with the red and blue currents. These indications would be correct if they were showing the currents indicated in red to be warmer than the surrounding waters and the blue currents to be colder than the surrounding waters. For example the Norwegian Current, although it is red, is a lot colder than any equatorial current which is marked black. The following map is more accurate in showing the actual water temperatures.

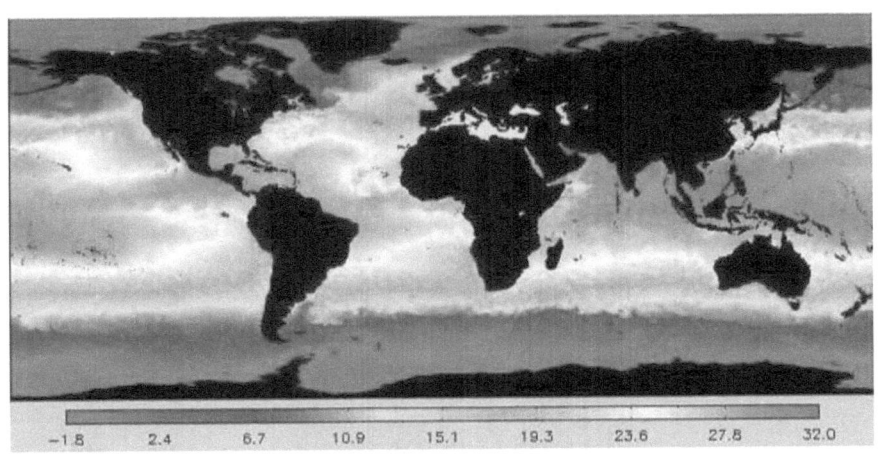

OCEANIC SURFACE WATER TEMPERATURES IN DEGREES F

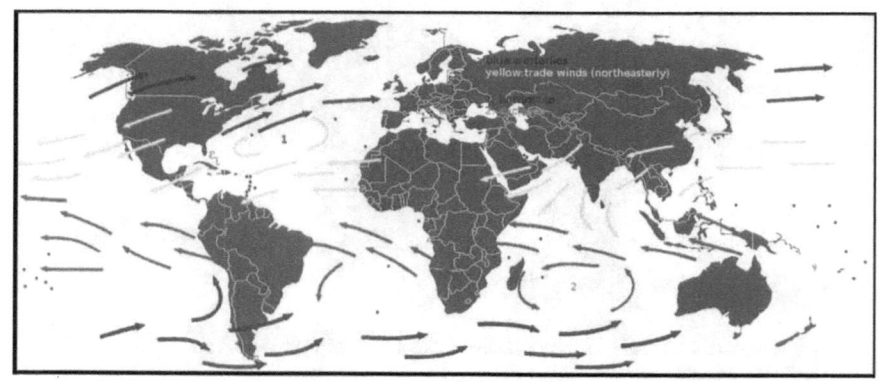

PREVAILING WINDS GLOBE MAP

This map from Wikipedia of trade winds is also in public domain and clearly shows the prevailing wind direction mainly in the tropics. I say this because the tropics are the engine of all winds and remain consistent. This is where the air mass is pushed in a westward direction from the sun and major weather disturbances are created. This mass air movement in the tropics is the cause of the mass air movement in the higher latitudes to move the opposite way. But the wind direction is not as consistent as it is in the tropics. So trade winds in higher latitudes are less predictable and their life is shorter to have a significant effect on sea currents.

If it was possible to open an exit in Mexico or in Panama for the currents to cross out into the Pacific then I can safely speculate that the Gulf Stream would die out and the weather in Northern Europe would change dramatically. I can also safely speculate that the surface in the Gulf of Mexico is considerably higher than the surface in the Pacific. Someday if I find the money and the time I will attempt to prove it. I hope someone will step forward and help me to do it.

My statement also more accurately explains the strong equatorial currents as they also move from east to west and the creation of the **counter equatorial currents** generated from the tropical storm winds on the side of the equator which blow in the opposite direction from west to east but are not as strong because they are only reinforced when a tropical storm is present. Tropical storms in the Pacific are more frequent and for this reason the counter equatorial current there is much stronger.

There are people who have studied the **Gulf Stream** and have concluded that it packs 100 times the energy we consume on the whole earth. If this is true, then the energy packed in **all the oceanic currents** has to be about 1,000 times the energy we consume on the earth. Now if the energy in all the oceanic currents is 1,000 times the energy we consume on the earth, how much larger does the energy of the wind which generates the currents have to be? Is 10,000 times a good figure? The wind, though, which generates the oceanic currents does not include the energy packed in the tropical storms and all the other storms around us. So if we include the energy of all the winds, it is safe to say that energy packed **in all the winds** is conservatively 100,000 times the energy we consume on the whole earth. Yet the **sun** is the one that generates all the winds and this as a byproduct of the energy with which the sun is bombarding the earth every day. Therefore, who controls our environment, man or the sun? I would say the sun.

We have seen how the moist air climbs up in the atmosphere and how the energy of the sun causes this air to rise and fall day after day in the tropics. The following statement may appear as an exaggeration, but believe me it is very accurate. **The effect the sun has in its travel over the tropics is in actuality an explosion within the atmosphere of astronomical proportion in its magnitude.** We do not realize it because it is not instantaneous as in a bomb exploding. It takes four to five hours to reach its highest expansion and then four to five more hours to settle. But its magnitude is many times larger than any known explosion. Plus this explosion is invisible until a tropical storm is developed to generate clouds, and then, when eventually as it becomes visible, we lose the perception that the sun is the one that has put it into motion.

This **rise and fall of the moist air is the engine which generates all the winds** especially the persistent ones from east to west in the tropics.

This **rise and fall of the moist air is also the engine which creates the strong disturbances in barometric pressures**. When tropical storms become hurricanes and typhoons they are extremely potent and cover large areas of the open sea, and in so doing, they effect the generation of more barometric disturbances farther away from the tropics. These barometric disturbances away from the tropics generate another classification of storms in the latitudes around forty degrees. **These storms are less predictable in their travel but move in a general west to east direction**.

Not only do these winds travel long distances, but they also cause the disturbances in atmospheric pressures as well. The effects become more dramatic when rain gets in the mix and, worse yet, when hail introduces its influence. All these destructive atmospheric phenomena from the North Pole to the South Pole and from Australia to Siberia and from Japan to Argentina have their source from one and only one thing, the energy from the sun. Yes, it is the sun that gives the energy and produces the winds, and the wind carries water for us to the mountains and snow to the poles. The heat evaporates water from the seas and protects the water from boiling. Yes, it is the sun that makes the winds which in turn cause the oceanic current to bring relief to northern Europe.

If you have any doubt about this matter you can experiment. At the side of a pool, place a small fan and watch what happens. You will see the small waves forming and then you will see the mass of the whole water in the pool start rotating following the air flow and coming around from the other side of the pool. If you want, you can use a more proportionally dimensioned pool, both in depth and width and with islands and other land formations, and you will see the Gulf Stream being formed before your eyes.

Most of the oceanographers do not even know of the existence of winds that blow in the same direction all year long especially in the tropics. And those who know it do not associate it with the Gulf Stream. Therefore, how would their thoughts go to this explanation about how the Gulf Stream is formed?

In the island of Aruba there is at the west end of the island a large refinery where a large smokestack is erected. The fumes emitted from that stack had an easily recognizable odor. Often our destination was Maracaibo and our route was a considerable distance west from Aruba. I remember sensing that odor from the Aruba smokestack and it was giving me a sure indication of our position in our approach to Venezuela.

But look, what else takes place in the Caribbean Sea. The warm sun is heating all the water on the surface of that sea. Warm water also floats on the surface. As the wind blows over the surface of the sea it skims this warm water and pushes it ever farther west into the Gulf of Mexico. Where does this mass of warm water go from there? Well, it attempts to escape north of Cuba and south of Florida by turning east. There it meets the North Equatorial Current coming west from the north side of Cuba. These currents, as they combine together to form an even stronger stream, have only one place to escape and that is by turning north next to the east coast of Florida.

We also spoke of "skimming" the warm surface water in the Caribbean. What takes place in the Maracaibo Lake? It is not really a lake but a gulf where tanker ships have gone in and out for many years before I was going there in the 1960's, and they still keep going even now, fifty years later.

This is what I remember about Maracaibo - a forest of oil wells in the middle of the sea. If there are any spills in the well-farms or the ships, they will be skimmed away from there by the wind and the currents and will hit Florida, Georgia, the Carolinas and maybe even England. Is there a problem of pollution in Maracaibo Lake? Absolutely there is, and locally, there are a lot of complaints. But somehow drilling in the Gulf of Mexico is the dangerous problem according to our government and environmentalists so we are not allowed to drill, but no one in the international media talks about the Maracaibo Lake problem where drilling has been going on for the last one hundred years. I just want to make a point for the environmentalists who also strive for the "global warming" danger.

Aside from the things we spoke about so far concerning the Gulf Stream, there is another issue. Correctly we know that hurricanes do intensify as when they travel over warmer surface waters of the sea. Skimming the warm water from the Caribbean Sea makes the Gulf of Mexico and the Gulf Stream much warmer which results in hurricanes like Katrina. But has anyone explained why this is the case? In some degree I hinted earlier. I did say that the warmer the temperature, the more intense the evaporation becomes. Add also that the water being warmer is closer to the evaporation level. As that heated up evaporated air expands to even higher levels in the atmosphere, the result is that this moisture-rich air eventually will go high enough where the temperature is a lot colder to become rain, and the tropical storm intensifies.

Finally I do agree that the Gulf Stream packs one hundred times the world's consumption of energy. What I want to make clear is that this humongous amount of energy in the Stream is completely unrelated to the energy consumption we mentioned so far about evaporation, rivers and glaciers and also to thunder and other environmental phenomena.

Therefore, **wind is the cause of the Gulf Stream** and the cause of all the other currents and streams in all the oceans. Well, I can qualify this better. There are some local isolated palindromic currents called tides caused by the moon's gravitational forces. More accurately, though it is the **sun which causes the wind**. Wind is also the cause of all the storms we have discussed. How can you estimate the energy packed in a storm like Katrina and all the other storms which occur constantly around us every day?

We have mentioned before that a larger portion of the sun's energy is used daily to warm up the air itself.

We have calculated the energy consumed in evaporation of water because we know the amount of rainfall.

Huge amounts of energy are also manifested as wasted energy as in floods, in tornados, in tropical and other storms or rivers which empty all the way to the sea and in electrical energy from lightning because we do not know how to control it and make good use of it. Yet, all these uncalculated forms of energy are manifested because of the sun.

This is what good scientists should be encouraged to do: To use their initiative, and their abilities to make use of some of this otherwise wasted energy the sun is giving us.

Sun's Energy Heating the Air

We know that on average the air at any place on the earth rises about 10 degrees during the day when the sun is out if there is some humidity in it, and then at night when the sun is not present, the temperature again on the average drops about 10 degrees.

This says that without the presence of the sun, the atmosphere of the earth is losing heating energy so we are lucky we have the sun in our proximity to send us warmth to make up for the losses during the night. Another way to see this effect is at the poles of the earth where the temperatures in the winter plunge to -40 and -50 degrees. How can we measure the energy we need to make up the loss of energy from the whole amount of the atmosphere?

Scientists have the knowhow, the computer power and the time to tell us, but for some reason avoid the responsibility to tell us. So the responsibility is left up to me and people like me to find the answer. I know that my answer is not going to be very accurate, nor is the scientist able to give an accurate answer. My intent in this discussion is to bring up a presentation we can grasp in our mind by cutting some corners.

We easily can refer to our homes, where we spend energy to maintain the air inside the house at a comfortable level. Our house has the same problem of heat losses especially in colder territories.

This is the reason we need a furnace because the house has losses of temperature. We try to minimize these losses by insulating the walls and making our windows and doors better, but we still lose energy. We need to make up the lost energy and we use a furnace.

The heating industry for homes and other buildings has devised a way to estimate the size of a furnace that is needed to install in a house. To do this they first take into account the local climate. What they look for in the local climate is how low the temperatures go in the winter. For example, if the minimum temperature is 5 degrees and we want to maintain the temperature inside at 65 degrees this gives us for our calculation a **temperature differential of 60 degrees**.

The **temperature differential for the atmosphere** is the total amount of energy to change the temperature of the air about **10 degrees**. All this energy is lost at night and regained during the day.

The industry also has developed tables which help to calculate the losses from building materials. The materials of concern here are the different materials used for doors, windows, exterior walls, and ceilings. These tables help the builder estimate the losses of energy in the house with the materials which are used and for a variety of temperature differentials.

I used a manual I have with the title "Warm Air Heating for Climate Control". In this book they use an example of a house which is typical and at a temperature differential of 60 degrees.

This typical house has 2,000 square feet at a 60 degree temperature differential.

Again the materials used are typical materials and for this example of a temperature differential of 60 degrees, the estimated heat losses were about **100,000 BTU/hr.** It is easy to understand that if the house was in a warmer place with a temperature differential of 30 degrees, the heat losses would be half as much and for this house it would be 50,000 BTU/hr.

This is telling us that in order to overcome the losses in this typical house, the furnace needs to provide a heating energy of 100,000 BTU/hr. We will convert this energy to 12 hours because this is how long the sun is out, making up the losses of heat. So the typical losses of the house in 12 hours are **1,200,000 BTU.** This is the amount of energy the furnace needs to provide if the temperature differential had remained at 60 degrees for all the 12 hours.

In this house we make up the losses by heating up the entire air volume within the house which here is 2,000 square feet times 10 feet of room height which gives a **1 house air volume of 20,000 cubic feet of air**.

Again because the temperature differential in the atmosphere is 10 degrees we will convert the losses in this house to correspond to a temperature differential of 10 degrees also. In the same house, if the temperature differential was half as much like 30 degrees, then the heat losses would be only 600,000 BTU. Therefore, the 10 degree temperature differential since it is 1/6 of the 60 degree temperature differential, the heat loss in this house is 1,200,000/6 = **200,000 BTU**, or **2 x 10^5 BTU** to make up the losses in **20,000 cubic feet** of air in 12 hours.

The earth's surface is 500,000,000 square kilometers = 5×10^8 square kilometers.

1 square kilometer is roughly 10^7 square feet.
Earth's surface is: **5×10^{15} square feet.**

Conservatively, I will take as the thickness of the atmosphere that the sun is heating to only 10,000 feet high. If I was to assume that thickness of the atmosphere is larger than the 10,000 feet then the result of my calculation would be greater in my favor. So, the volume of the atmosphere the sun is heating is: **5×10^{19} cubic feet.**

Dividing this volume with 20,000, the volume of air in one house, this gives us 2.5×10^{15} or

2,500,000,000,000,000   House volumes

The sun's heating energy in the atmosphere is: $2.5 \times 10^{15} \times 2 \times 10^5$
= 5×10^{20} BTU/day
= 5×10^5 Exajoules/day
The sun is sending in just one day more than 1,000 times the world's energy consumption in one year.
Roughly it is 400,000 times the world's energy consumption in one year.

Just based on this, when the sun is sending down to earth this staggering amount of energy, what difference would it make if we even stopped traveling, stopped heating our homes and stopped driving our cars. Eliminating all our energy consumption on the earth, in all practicality, makes no difference whatsoever to our environment.

Sun's Energy Manifestations

We have discussed a variety of such manifestations of the sun's energy. We did this to draw attention in different ways because not all people have the same experiences or the same background in life. And because all these manifestations are related to the sun's energy, in this section I will make an attempt to list them and in some way differentiate them.

1. **Atmospheric heating**

   400,000 times the world's energy consumption.
   The source is the sun.

2. **Oceanic water heating**

   Amount of energy is not calculated. Primary source is again the sun. It could be estimated if we had the amount of water which is traveling in the equatorial currents, and if we estimate how much the water increases in temperature while it travels in the tropics.

3. **Evaporation**

   The source is water's property as a refrigerant. The primary source is the sun. It packs 1,500 times the world's energy. Even this is insignificant compared to item 1.

4. **Rivers, glaciers, falls**

   This energy is not calculated. The source is evaporation and wind. Primary source is the sun. We make a minimum use of their energy.

5. **Winds, storms, monsoons, hurricanes, tornados, etc**

This energy is also not calculated. Primary source is the sun. We can make a lot more use of its energy.

6. **Electrical energy, thunder and lighting**

This energy is also not calculated. Source is moisture from evaporation. Primary source is the sun.

7. **Oceanic currents**

This energy is partially calculated. All oceanic currents are caused from winds. Primary source is the sun.

8. **Oceanic waves**

These waves hit all the coasts around the world. How can we calculate the energy they can produce? Winds produce the waves, and as such, still, the sun is the primary source.

9. **Light**

Definitely from the sun.

This is the best example of abandoned energy taken for granted. We do turn the night into day in a confined space like a stadium. But do we have the energy to do the same thing for a whole city? How about a whole state? How about the whole country? How about the whole American Continent?

I have been involved with these issues for a long time even before global warming had become such a serious topic. I have seen the destructive effect a man has on the environment. I have seen it in the Mediterranean Sea and in Maracaibo and I have heard of the damage we did in the fishing industry in Chesapeake Bay. I do not know about Maracaibo Lake, but the Mediterranean Sea and Chesapeake Bay now are clean. In many instances because of our ignorance and our greed, we have done these things. Yes, we can make a mess in our house, our neighborhood, and our rivers, and it is the responsibility of all people to keep these places clean always. But it is a sign of a proud man to think that he is strong enough to damage the earth's environment. We do not need an industry to keep these places clean; we all need to do it for free. A man is no more than an ant in front of our oceans and our mountains. Those who make the global warming statements need to put forth the effort to explain where the danger is and how. The amount of energy to warm up the air 10 degrees alone is just staggering and this is not all. Everything under the sun is receiving this astronomical size of energy, not just the air of the atmosphere. If we did not have this amount of water on the earth, deserts would be more prevalent and our life more difficult. Water counteracts to a large degree the warming effect the sun has on earth. The oceanic warm currents of the tropics do the same thing in moderating the climate. As these currents move away from the tropics they lose a lot of the temperature they have acquired from the sun in the evaporation of their warm water and coming in contact with colder winds and colder currents coming from the arctic seas. The arctic areas are getting very little sun, and any moisture which travels that far, freezes and comes down on the ground.

My final point is this: I agree that we all should put the effort to be frugal, both rich and poor. The energy from the ground like oil is less expensive and this is the reason we use it. Curtailing the energy production in our country only results in higher prices, the economy goes down, and we mortgage our country to other nations. If you let businesses handle their economy, everything is taken care of because people who work for their needs, even their wants, do not want to lose what they have. The only ones who do not appreciate their income are the thieves, the government contractors, and the welfare recipients. When entrepreneurs are rewarded for their innovations, things are happening. In these days the only innovators who are rewarded are the Chinese because they can produce not only their innovations but also our innovations cheaper.

In any country and in any political system, if you do not reward those who invent and produce, you are destined to lose them both – the producers and the innovators - and in turn everyone loses.

I will give an example.

You have two independent companies. They produce the same product, they have similar assets, and hire the same number of people. One company makes a profit of $150,000 and the other, because of streamlining their operation, makes a $250,000 profit. When you tax the profitable company with a higher tax rate, all you do is penalize that company for their productivity and their innovation. I would like to see the productive company rewarded for their innovations and their streamlining. I would even introduce some formula which would evaluate the productivity and give them better tax rates for hiring more people. Then you will see a competition among productive companies to make their products cheaper. And still, most productive companies would be as happy not to be penalized with higher tax rates.

Most of our businesses would be happy to know that the next year they will have the same expenses as they had the current year, even if their profits are not as great. They can live with it. But when they are faced with higher taxes or bigger regulations, they become insecure in their ability to make ends meet and reduce their operation, fearing that in the future they may lose it all.

In our country we should promote scientists and innovators to find ways to use the resources our sun is giving us for free. With some promise to keep what they earn and better yet with some incentive to improve their product, people will be more inclined to try something new. Any small success will inspire them for even bigger success. Only with a flat and low tax rate for all the working individuals will the sky be the limit. Today we hear in the news that more than half of the people do not pay any taxes, and many of them get tax credits. This is absurd. It is equally absurd saying that climate changes are caused by our consumption of energy, while the sun is literally bombarding our planet with one million times more energy than the puny amount we consume.

THE END

Notes

Notes

Notes

Notes

Notes

Notes

Notes

Notes

Notes

Notes

Notes

Notes

Notes

Notes

Notes

Notes

Notes

Notes

Notes

Notes

Notes

Notes

Notes

Notes

Notes

Notes

Notes

Notes

Notes

Notes

Notes

Notes